6th Grade Math

Volume 2

© 2013 OnBoard Academics, Inc
Newburyport, MA 01950
800-596-3175

www.onboardacademics.com
ISBN: 978-1494857295

Table of Contents

Adding & Subtracting Fractions (with unlike denominators)   4

Adding & Subtracting Fractions (with unlike denominators) Quiz   9

Adding & Subtracting Mixed Numbers (with unlike denominators)   10

Adding & Subtracting Mixed Numbers (with unlike denominators) Quiz   16

Multiplying Mixed Numbers   17

Multiplying Mixed Numbers Quiz   22

Dividing Fractions & Mixed Numbers   23

Dividing Fractions & Mixed Numbers Quiz   30

Adding & Subtracting Fractions
(with unlike denominators)

Key Vocabulary

denominator

Least Common Denominator (LCD)

Model adding fractions with unlike denominators.

Shade in the sum to find the answer.

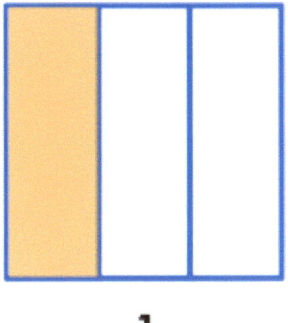
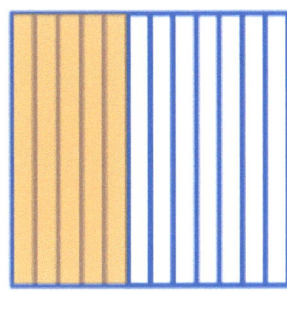
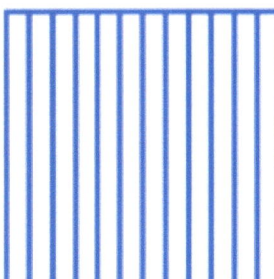

$$\frac{1}{3} \quad + \quad \frac{5}{12} \quad =$$

$$=$$

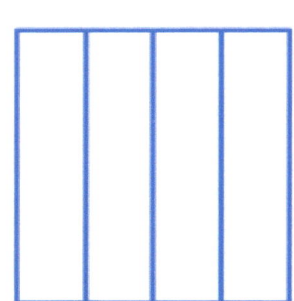

Model subtracting fractions with unlike denominators.

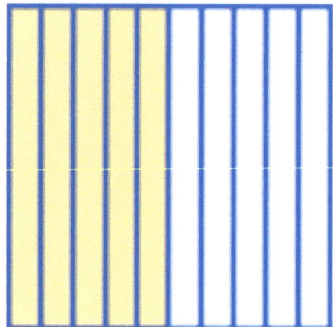

$$\frac{1}{2} \; - \; \frac{2}{5} \; = $$

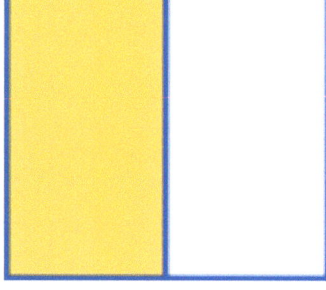

Practice adding and subtracting fractions.

$$\frac{11}{12} - \frac{7}{10} = \frac{55}{\boxed{}} - \frac{42}{\boxed{}} = \frac{13}{\boxed{}}$$

$$\frac{5}{8} - \frac{5}{14} = \frac{\boxed{}}{56} - \frac{20}{56} = \frac{\boxed{}}{56}$$

$$\frac{5}{6} - \frac{3}{4} + \frac{7}{12} = \frac{10}{\boxed{}} - \frac{9}{\boxed{}} + \frac{7}{\boxed{}} = \frac{\boxed{}}{\boxed{}} = \frac{\boxed{}}{\boxed{}}$$

Stretch your knowledge.

Dividing Michael's Salary

Michael spent $\frac{1}{4}$ of his salary on clothes, $\frac{2}{5}$ on his cell phone, and $\frac{3}{10}$ on going out with his friends.

What fraction of his salary does he have left?

Name_____

Adding & Subtracting Fractions (with unlike denominators) Quiz

1 True or false: $\frac{1}{4} + \frac{2}{5} + \frac{3}{10} > 1$?

2 $\frac{1}{3} + \frac{5}{6} =$

   A $\frac{11}{6}$

   B $1\frac{1}{6}$

   C $\frac{7}{6}$

   D $1\frac{1}{3}$

3 What is the LCD for $\frac{3}{5}$ and $\frac{4}{8}$?

4 Jack spends $\frac{2}{5}$ of his allowance on music and $\frac{3}{10}$ on candy. How much does he have left?

Adding & Subtracting Mixed Numbers (with unlike denominators)

Key Vocabulary

mixed numbers

Least Common Denominator (LCD)

Study the model for adding mixed numbers.
Fill in the answer and then simplify.

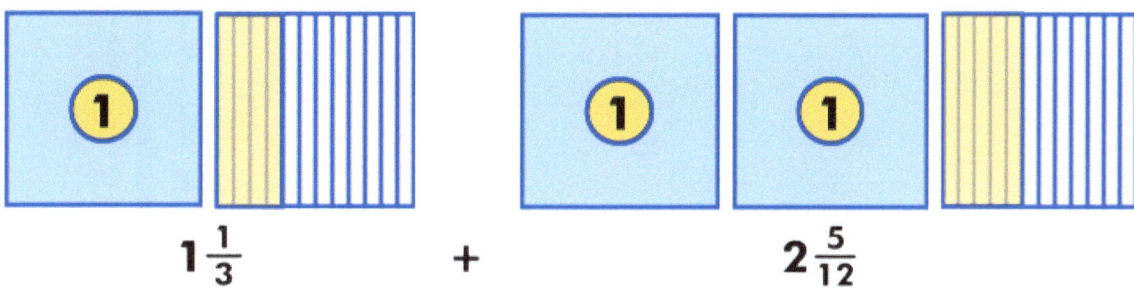

$$1\frac{1}{3} \quad + \quad 2\frac{5}{12}$$

$$= 3\frac{9}{12} \qquad 3\frac{3}{4}$$

Model and complete this problem.

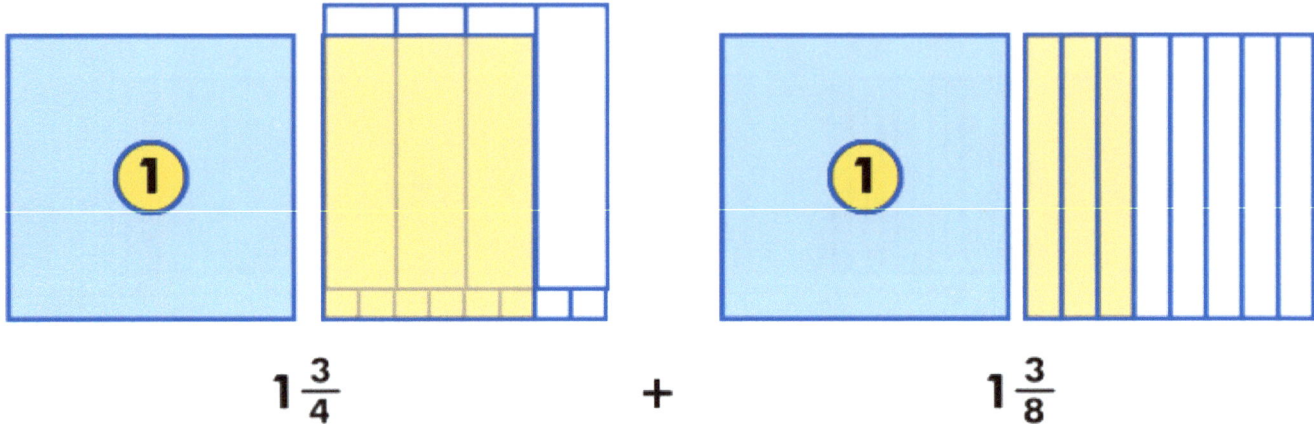

$$1\frac{3}{4} \qquad + \qquad 1\frac{3}{8}$$

=

Model subtracting mixed numbers.
Fill in the answer.

$$2\frac{1}{2} \quad - \quad 1\frac{3}{4}$$

Study the illustration below to learn to rewrite the fraction part of mixed numbers with a common denominator.

$$4\frac{3}{4} \;-\; 1\frac{5}{6}$$

$$= \left(4\frac{9}{12}\right) - 1\frac{10}{12}$$

$$= \left(3\frac{21}{12}\right) - 1\frac{10}{12}$$

$$= 2\frac{11}{12}$$

Study the illustration below to learn to write mixed numbers as improper fractions with common denominators.

$$4\frac{3}{4} - 1\frac{5}{6}$$

$$= 4\frac{9}{12} - 1\frac{10}{12}$$

$$= \frac{57}{12} - \frac{22}{12}$$

$$= \frac{35}{12}$$

$$= 2\frac{11}{12}$$

Name_____

Adding & Subtracting Mixed Numbers (with unlike denominators) Quiz

(1) $1\frac{2}{5} + 4\frac{3}{10} = 5\frac{7}{10}$ True or false?

(2) $3\frac{2}{3} + 1\frac{3}{4}$ can be written as:

   A $3\frac{4}{6} + 1\frac{5}{6}$

   B $3\frac{8}{12} + 1\frac{9}{12}$

   C $3\frac{6}{12} + 1\frac{9}{12}$

   D $3\frac{4}{6} + 1\frac{3}{6}$

(3)   What is $12\frac{3}{4} - 4\frac{5}{12}$ written in simplest form?

(4) $3\frac{5}{6} + 4\frac{1}{3} + \frac{3}{4} = 8\frac{x}{12}$ x = ?

Multiplying Mixed Numbers

Key Vocabulary

mixed number

improper fraction

The area of a rectangle.
Study the illustrations below to learn how mixed numbers are used when finding the area of a triangle.

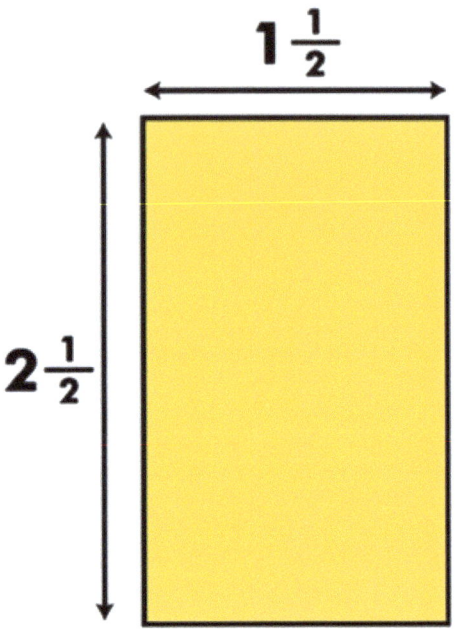

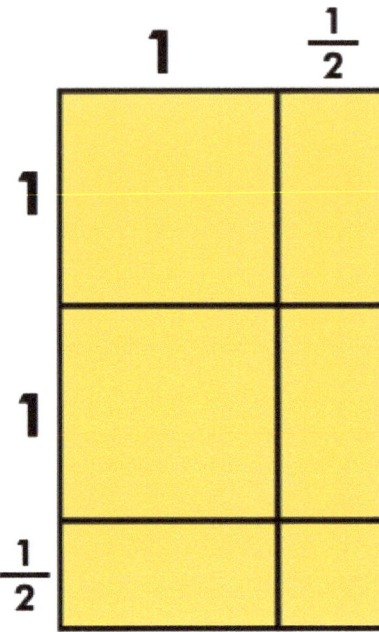

$$1\frac{1}{2} \ \text{x} \ \ 2\frac{1}{2}$$

1 **Write as improper fractions**		$\frac{3}{2} \ \text{x} \ \frac{5}{2}$
2 **Multiply**		$\frac{3 \times 5}{2 \times 2} = \frac{15}{4}$
3 **Simplify if necessary**		$\frac{15}{4} = 3\frac{3}{4}$

Multiplying a mixed number by a whole number.
Study the illustration below to learn to multiply a mixed number by a whole number.

$$3 \quad \times \quad 2\frac{1}{2}$$

$$\frac{3}{1} \quad \times \quad \frac{5}{2} \; = \; \frac{15}{2} \; = \; 7\frac{1}{2}$$

Write the whole number in fraction form, multiply and then simplify.

Simplifying before multiplying can make the problem easier to handle.

$$1\frac{1}{11} \quad \times \quad 3\frac{3}{4}$$

① **Write as improper fractions**		$\frac{12}{11} \times \frac{15}{4}$
② **Simplify**		$\frac{\overset{3}{\cancel{12}}}{11} \times \frac{15}{\underset{1}{\cancel{4}}}$
③ **Multiply**		$\frac{3 \times 15}{11 \times 1} = \frac{45}{11}$
④ **Simplify**		$\frac{45}{11} = 4\frac{1}{11}$

Practice Multiplying Fractions

$$1\frac{1}{4} \times \frac{5}{6} \qquad \frac{\Box}{4} \times \frac{\Box}{6} = \frac{25}{\Box} = \frac{\Box}{\rule{1cm}{0.4pt}}$$

$$2\frac{1}{5} \times 1\frac{4}{7} \qquad \frac{\Box}{5} \times \frac{\Box}{7} = \frac{\Box}{\Box} = \frac{\Box}{\rule{1cm}{0.4pt}}$$

Practice multiplying fraction. Simplify first!

$$4 \times 2\frac{7}{8}$$

$$\frac{\square}{1} \times \frac{23}{8} = \frac{\square}{2} = \square\frac{1}{2}$$

$$2\frac{2}{5} \times 3\frac{1}{3}$$

$$\frac{\square\;12}{15} \times \frac{210}{3} = \frac{\square}{1} = \square$$

Name_____

Multiplying Mixed Numbers Quiz

(1) **True or false?** $12\frac{4}{7} = \frac{16}{7}$

(2) **What is the area of the rug?**

    A $15\frac{5}{16}$ ft^2

    B $12\frac{3}{8}$ ft^2

    C $12\frac{3}{16}$ ft^2

    D $7\frac{4}{10}$ ft^2

$4\frac{3}{8}$ ft

$3\frac{1}{2}$ ft

(3) $2\frac{2}{5}$ x $1\frac{3}{7}$ = ?

(4) **What is the missing numerator?** $4 \times \frac{?}{5} = 1\frac{3}{5}$

Dividing Fractions & Mixed Numbers

Key Vocabulary

reciprocal

Dividing a Fraction

$$2 \div \frac{1}{3} =$$

| $\frac{1}{3}$ | $\frac{1}{3}$ | $\frac{1}{3}$ | | $\frac{1}{3}$ | $\frac{1}{3}$ | $\frac{1}{3}$ |

| 1 | | 1 |

Using Reciprocals to divide Fractions

Two numbers whose product is 1 are called *multiplicative inverses* or *reciprocals*.

$$\frac{1}{3} \times \frac{3}{1} = 1$$

Dividing by a fraction is the same as multiplying by the reciprocal of the fraction.

$$2 \div \frac{1}{3} = 6 \qquad \frac{3}{1}$$

$$2 \times \frac{3}{1} = 6 \qquad \frac{1}{3}$$

Reciprocal

What is the recriprocal of 4?

Hint $4 = \frac{4}{1}$

Solve using reciprocal fractions.

$$\frac{5}{7} \div \frac{2}{3}$$

$$= \frac{5}{7} \times \boxed{}$$

$$= \boxed{}$$

$$= \boxed{}$$

$$6 \div \frac{3}{4}$$

$$= \boxed{} \times \boxed{}$$

$$= \boxed{}$$

$$= \boxed{}$$

Dividing Mixed Numbers Using Reciprocal Fractions
Study the illustration below.

$$1\frac{3}{4} \div 2\frac{1}{8}$$

$= \frac{7}{4} \div \frac{17}{8}$ **Write mixed number as improper fraction**

$= \frac{7}{4} \times \frac{8}{17}$ **Multiply by the reciprocal**

$= \frac{7}{\cancel{4}^{1}} \times \frac{\cancel{8}^{2}}{17}$ **Simplify if necessary**

$= \frac{14}{17}$ **Solution**

Complete these division problems.

First select the proper equation form the four options. Write the proper equation in the box with the dotted lines then solve.

$$3 \div 2\frac{1}{3}$$

$$\frac{3}{1} \times \frac{3}{7} \qquad \frac{1}{3} \times \frac{3}{7}$$

$$\frac{1}{3} \times \frac{7}{3} \qquad \frac{3}{1} \times \frac{7}{3}$$

$$1\frac{5}{6} \div \frac{1}{2}$$

$$\frac{15}{6} \times \frac{2}{1} \qquad \frac{11}{6} \times \frac{2}{1}$$

$$\frac{15}{6} \times \frac{1}{2} \qquad \frac{6}{11} \times \frac{2}{1}$$

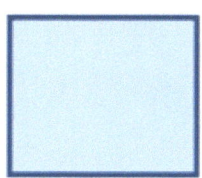 =   =

Chew over this division problem......

A group of friends purchase six sub sandwiches so that they can all have $\frac{2}{3}$ of a sandwich each.

How many friends are there?

Stretch your knowledge.

How much fabric is needed for each dress?

 Ashima's mother is making some dresses for a family wedding. She has 22 yards of fabric which is enough to make 6 dresses with $\frac{1}{4}$ yd left over.

How much fabric does she need for each dress?

Name_____

Dividing Fractions & Mixed Numbers Quiz

1 True or false: if each student eats $\frac{1}{4}$ of a pizza, 34 students will eat $8\frac{1}{2}$ pizzas?

2 $2\frac{3}{8} \div 1\frac{3}{5} =$

   A $\frac{8}{19} \times \frac{8}{5}$

   B $\frac{23}{8} \div \frac{13}{5}$

   C $\frac{19}{8} \times \frac{5}{8}$

   D $\frac{8}{23} \times \frac{5}{13}$

3 What is the missing denominator? $\frac{3}{7} \div 1\frac{1}{4} = \frac{12}{?}$

4 If you need $2\frac{1}{4}$ cups of flour to make 12 cup cakes, how many cup cakes can you make with $15\frac{3}{4}$ cups of flour?

Newburyport, MA 01950

1-800-596-3175

OnBoard Academics employs teachers to make lessons for teachers! We create and publish a wide range of aligned lessons in math, science and ELA for use on most EdTech devices including whiteboard, tablets, computers and pdfs for printing.

All of our lessons are aligned to the common core, the Next Generation Science Standards and all state standards.

If you like our products please visit our website for information on individual lessons, teachers licenses, building licenses, district licenses and subscriptions.

Thank you for using OnBoard Academic products.